MW01293361

WAITING for WINTER

A True Story of Resilience and Joy in the Face of Climate Change

by Stephen Gorman

CAPSTONE EDITIONS
a capstone imprint

In winter, the Arctic is a world of white.

The vast blue ocean freezes
and turns into a solid, snow-covered surface.

When the ocean freezes,
polar bears venture out on the ice to hunt.

They wander the surface of the frozen Arctic Ocean
looking for ringed seals, their favorite food.

Although polar bears are great swimmers, they can't swim nearly fast enough to catch seals in open water.

The bears can only catch seals by grabbing them at their breathing holes in the sea ice.

In the heart of winter, the Arctic is still white and frozen.

But as Earth warms and the climate changes, the sea ice melts earlier in the spring and forms later in the fall.

When there is no ice, the polar bears can't hunt seals.
They are stranded on shore as they wait for the ocean to freeze.

And so, polar bears like these wander the shoreline
for months, hungry and waiting for winter.

In late fall, when the sea
should be frozen, these cubs in
Alaska's Arctic National Wildlife Refuge
have waited for winter for so long
their white fur coats have turned brown
from lying around in the sand.

And so, they wait . . .

and wait.

This cub waits for winter by playing with a toy that she made out of seaweed.

And one cub waits for winter by playing with a stick that looks just like a flute.

These cubs wait for winter by wrestling and roughhousing while their exhausted mother takes a little break.

But eventually, all the cubs grow hungry and tired of waiting for winter. They haven't had a good meal in months! And so, they take a lot of naps.

Even the grown-up bears are bored.

Eventually, the hungry cubs decide to find something to eat to tide them over while they wait. The cubs leave the beach and follow their mother into the open water.

They swim over to visit their human neighbors in Kaktovik, Alaska.
This Iñupiat Eskimo village sits on Barter Island just across the lagoon.

The people of Kaktovik catch bowhead whales in the fall. These giant whales migrate past the village on their way south to the Bering Sea, where they spend the winter.

The cubs know they will find leftovers from the Iñupiat whale hunt at the edge of town.

So, after lunch, the cubs play while their mothers watch.
The cubs spend hours scrambling over the giant whale bones.

Late in the day, there is a new chill in the air. As the cubs play in the ocean, they seem to ask one another, "Do you think winter is *finally* coming?"

The bears on shore seem excited too. Maybe they can sense a big change in the weather coming. Will the sea ice form soon?

Will they be able to get off this beach and hunt seals again?

Gently at first,
the snowflakes start to fall and keep falling.
Soon, there are several inches of fluffy white snow on the sand.

One of the cubs rolls in the snow with joy.

And then all the happy cubs begin to play in the new snow.

While their cubs play together in the snow,
the polar bear mothers gather at the edge
of the sea, watching it begin to freeze.

Soon, all the polar bears will get off the beach,
go out onto the sea ice, and find seals.
For at least one more winter they will wander
the frozen surface of the Arctic Ocean.

And so, for the polar bear cubs
of the Arctic National Wildlife Refuge in Alaska,
the long wait for winter is finally over!

The Arctic Region and the Arctic National Wildlife Refuge

The Arctic is located at the northernmost part of our planet. Scientists usually define the Arctic as the area within the Arctic Circle. This imaginary line circles the top of the globe.

The Arctic includes the Arctic Ocean and parts of Russia, Finland, Sweden, Norway, Iceland, Greenland, Canada, and the United States.

The Arctic National Wildlife Refuge is the largest National Wildlife Refuge in the United States. It is home to polar bears, caribou, wolves, and many other animals. Millions of birds from around the world migrate to the refuge each year.

The refuge is located on the traditional homelands of the Iñupiat and Gwich´in peoples. These Indigenous peoples continue to use the refuge for hunting, fishing, and gathering food.

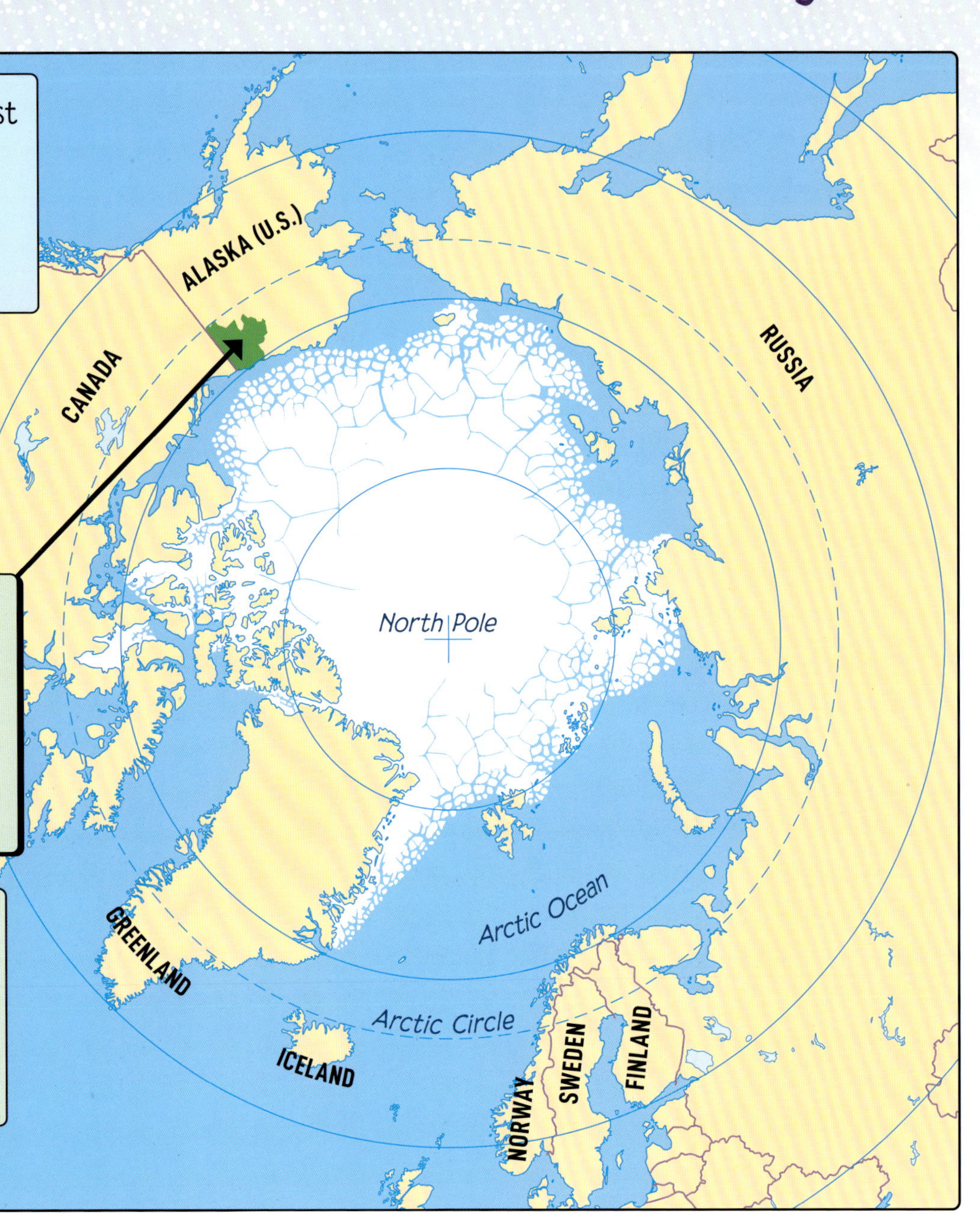

More About Polar Bears

Polar bears live along the shoreline and on the sea ice in the Arctic. When sea ice forms over the ocean in winter, most polar bears head out onto it to hunt.

Polar bears primarily hunt and eat seals, their favorite food. Polar bears often rest silently at a seal's breathing hole in the ice, waiting for a seal to come to the surface.

Polar bears are the biggest bears in the world. Large male bears can weigh nearly 2,000 pounds (900 kilograms). That's almost as much as a small car!

Polar bears are very strong swimmers. They use their large front paws, which are slightly webbed, to paddle forward through the water. Polar bears have been seen swimming hundreds of miles from land.

The changing climate is making it harder for polar bears to hunt. Sea ice melts earlier in the summer and forms later in the fall than it has in the past. Without the sea ice, polar bears must scavenge for other, less nutritious, food.

A Note to Caregivers: How to Talk to Children About Climate Change

We live in a time of great transformation, and young people already know more about climate change than you might think. Research shows that many children have already developed feelings of anxiety due to what they have overheard in the media or in adult conversations. As a result, their levels of emotional distress may be quite high. The problem is kids aren't getting their information about climate change from a trusted and reliable source—you.

Because children under six may be too young to understand climate change, leave out the facts and data when you talk with them about it. Instead, help them cultivate environmental awareness and a love of nature through outdoor play, information about animals and the changing seasons, an appreciation of natural beauty, and an understanding of the need to protect and preserve nature. Picture books like *Waiting For Winter* help young kids develop a love of the natural world that will last a lifetime and help them grow into good environmental stewards.

Kids aged seven to twelve are ready to hear about the scientific facts and data. At this age, kids are beginning to understand the larger implications of climate change and are making connections between changes in the environment and their effects on people and wildlife. Each time they read *Waiting For Winter*, kids in this age group will understand more and more about the impacts of climate change on the Arctic, the polar bears, the bowhead whales, and the Iñupiat. As these kids get older, their growing understanding will give rise to increasingly complex feelings and emotions. This is a good time to help kids identify and name those feelings and help them practice emotional resilience.

Many kids worry about how climate change will affect their lives as they grow up. They are scared, and they worry about their future. These feelings of fear are normal, healthy responses to the threats posed by the escalating climate emergency. However, it is important to remember that, unlike their parents, this is the only world children have ever known. They may be more comfortable than their elders with making sacrifices to their lifestyles to improve their world. In fact, while the climate crisis brings risks, it may also offer kids opportunities for empowerment.

Climate Change Conversation Starters

* The planet may change a lot during your lifetime. How does that make you feel?
* Nature is our home. How has it made your life better? What would you miss if something in nature was no longer here?
* What can you learn at school about Earth and nature to help you make sense of the changes in the environment?
* Where can you learn more about climate change outside of school?
* What types of changes can you make in your daily life to have a positive impact on Earth's climate?
* What skills can you learn to help your community in this challenging time of climate change?

Photo courtesy of Stephen Gorman

Stephen Gorman is an internationally recognized wildlife photographer and best-selling author. His work focuses on how cultural values and national mythologies shape our relationships to the world we live in and the diverse societies with which we share it. Gorman is the author and photographer of several books, including *The American Wilderness: Journeys into Distant and Historic Landscapes* and *Northeastern Wilds: Journeys of Discovery in the Northern Forest*. He also won the Benjamin Franklin Award for *Arctic Visions: Encounters at the Top of the World*, a book commissioned by the Inuit of Nunavik in Canada. Throughout his career, Gorman has worked on cultural and environmental assignments for leading periodicals such as *National Geographic*, *Discovery Channel*, and *Sierra*. His most recent exhibitions include *Down to the Bone*, a collaboration with beloved *New Yorker* artist Edward Koren at the Peabody Essex Museum in Salem, Massachusetts, and *Visions of Inuit Life–Photographs by Stephen Gorman* at the Museo del Oro in Bogota, Colombia. For more about Stephen's work, visit stephengorman.com.

Published by Capstone Editions, an imprint of Capstone
1710 Roe Crest Drive, North Mankato, Minnesota 56003
capstonepub.com

Library of Congress Cataloging-in-Publication Data is available on the Library of Congress website.

ISBN: 9781630793210 (hardcover)
ISBN: 9781630793227 (ebook PDF)

Summary: Journey to the Arctic, where polar bears must await the arrival of winter. With stunning photographs from award-winning professional wildlife photographer and author Stephen Gorman, *Waiting for Winter* tells the story of a group of polar bears living in Alaska's Arctic National Wildlife Refuge. Each year, these bears must wait longer and longer for winter, the season that brings them sustenance for the rest of the year. As we watch them await winter's snow and ice, we witness their challenges, resilience, and moments of joy in this time of climate change.

Image Credits
Getty Images: Dimitris66, cover and throughout (snowfall), PeterHermesFurian, 36

Designed by Sarah Bennett

Printed and bound in China. 6275